I Am An Old Horned Toad

I Am An Old
HORNED TOAD

by Lane Johnson

Illustrated by Jessica Helm
Music by Merle Travis

Brown Wing Press

Learn more about the book and the music online, at **azhornedtoad.com**

ISBN: 978-0-9844912-9-2

Brown Wing Press
Iowa City, IA

brownwingpress.com

Second Brown Wing Press Paperback Edition

BROWN WING
PRESS

FORWARD

When I was a little boy growing up in Kentucky I was always fascinated by ads in the back of magazines and comic books for mail order "Horned Toads from Texas." Well, that certainly interested me, but my bedroom was already full of local snakes, lizards and turtles and the idea of getting an animal in the mail seemed strange to me even then, so instead, I read everything I could find about horned toads and found that they are lizards not toads, and they looked so cool!

How was I to know then, that only a few years later, at the age of nineteen, I would begin a whole new life in the desert southwest, home to quite a few species of horned lizards?

Oh my gosh, I learned so much about this interesting lizard, but I'm not going to give any of it away, because it's all in this beautiful little book and all of it is true!

Over the years, I've come to think that all the creatures that live in the southwest borderlands have a story to tell or a song that needs to be sung and you hold in your hands the proof.

I sure hope you love this book as much as I do. Now turn the next page and get ready to sing as you read the story of that old horned toad.

Petey Mesquitey
Pearce, Arizona

Howdy and Thanks!

This little book has been about 10 years in the making. It has been a labor of love with a lot of eagle eyes, helping hands, and encouragement. Thanks mostly to my lovely bride, Nancy, who has gently persisted in keeping me at it. Thanks to Danny and Cassidy for singing and staying with me through the process. A big ol' thanks to the Tortolita Gutpluckers for always playing along. Thanks to the Stephen Clan for their perennial presence and support. Thankee' kindly to Diamond Jim and Lynn Hewitt for the recording and backing vocals. Jessica Helm was the artist who finally stuck with it to finish the beautiful illustrations. Thanks to Merle Travis for writing such a beautiful song, and to his daughter Merlene, for helping get permission to use it. Thanks to my longtime buddy, Petey Mesquitey for the forward. Thanks to Brown Wing Studio for putting it together, and to the moms, Leila and Stephanie, for their ideas. It took a whole herd of us to get it done, partners, but here it is! Hope you like it as much as we enjoyed putting it together.

Happy Trails,

Lane "Mad Doc" Johnson

July, 2011
Tucson, Arizona

More about the book and the music online, at **azhornedtoad.com**

sing along at azhornedtoad.com

I am an old horned toad
I'm as tough as can be
I live in the desert
Where there's barely a tree.
There isn't much water.
The sun here is hot.
I haven't got much
But I like what I've got.

Chorus

I'm as flat as a pancake
I'm sharp as a tack
I'm covered with horns
From my head to my back
I live with the cactus
All covered with spines.
So leave me alone
In this desert of mine

There's nothing I like more
Than to eat ants all day.
On a rock doing push-ups
Is my favorite play.

And if you rub my belly
I just might close my eyes.
And take a siesta
If I so decide.

Chorus

I'm as flat as a pancake
I'm sharp as a tack
I'm covered with horns
From my head to my back
I live with the cactus
All covered with spines.
So leave me alone
In this desert of mine

Well there's lots of critters
Who'd like to eat me.
But if'n they try it
They soon let me be.

'Cause when they bite down
They get quite a surprise
I stick out my horns
And shoot blood from my eyes!

Chorus

I'm as flat as a pancake
I'm sharp as a tack
I'm covered with horns
From my head to my back
I live with the cactus
All covered with spines.
So leave me alone
In this desert of mine

Oh, it's not very easy
To be a horned toad.
My home's been destroyed
By your buildings and roads.

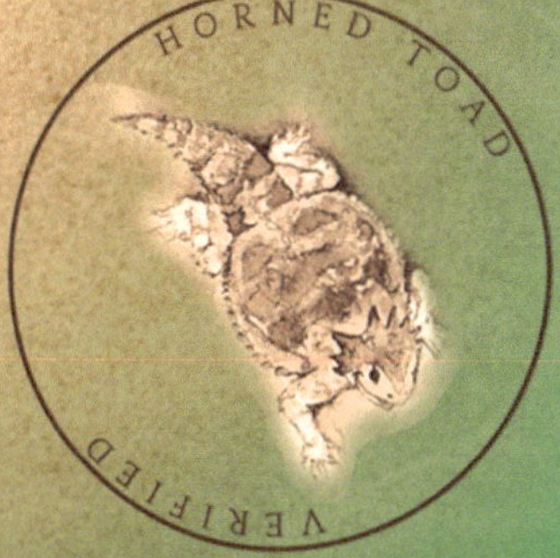

And that's why I took time
To sing you this song.
If you don't pay attention
You'll soon find that I'm gone!

Chorus

I'm as flat as a pancake
I'm sharp as a tack
I'm covered with horns
From my head to my back
I live with the cactus
All covered with spines.
So leave me alone
In this desert of mine

I Am An Old Horned Toad

(G) I am an old horned toad,
I'm (C) tough as can (D) be
I (G) live in the desert
where there's (C) barely a (G) tree.
There (G) isn't much water,
the (C) sun here is (D) hot.
I (G) haven't got much,
but I (C) like what I've (G) got.

CHORUS:
I'm (D) flat as a pancake,
I'm (C) sharp as a (G) tack
I'm (D) covered with horns
from my (C) head to my (G) back
I (G) live with the cactus
all (C) covered with spines,
So (G)leave me alone
in this (C) desert of (G) mine.

There's nothing I like more
than to eat ants all day.
On a rock doing push-ups
is my favorite play.

And if you rub my belly
I just might close my eyes,
and take a siesta
if I so decide.

Chorus

Well there's lots of critters
who'd like to eat me,
but if'n they try it
they soon let me be.

'Cause when they bite down
they get quite a surprise;
I stick out my horns
and shoot blood from my eyes!

Chorus

Oh, it's not very easy
to be a horned toad.
My home's been destroyed
by your buildings and roads.

And that's why I took time
to sing you this song;
If you don't pay attention
you'll soon find that I'm gone!

A Little More About Horned Toads

Horned toads are not really toads at all, they're lizards. There are seven different types of horned toads that live in the Western part of the United States.

Horned toads are slow moving. They depend on their coloring to help disguise them from danger. They can puff themselves up to make their horns point straight out if an animal tries to eat them. And yes, they can actually shoot blood from their eyes!

Prehistoric Native Americans also enjoyed Horned Toads. Petroglyphs, pottery, and baskets commonly used Horned Toad themes. Older Navaho people refer to Horned Toads as "Grandmother" or "Grandfather." They do appear very old and wise. They have survived in the desert a long time.

Horned Toads are gentle creatures. If you carefully catch one and rub its belly, it may take a little nap. Horned Toads eat mostly ants, a few other bugs, and some plants.

Horned Toads are disappearing because of construction and collecting. Horned Toads rarely survive more than a few weeks if they are captured. It is best to enjoy Horned Toads in the outdoors and then leave them there!

For more information about Horned Toads, visit our website: **azhornedtoad.com**

About

AUTHOR AND LYRICS:

Lane "Mad Doc" Johnson

Lane Johnson is a "near native" of Arizona, growing up in Prescott and Tucson, Arizona. In addition to playing with horned toads most of his life, he has played guitar in a "Brown Thorn" (Arizona Blue Grass) band, the Tortolita Gutpluckers, for over thirty years. You can visit the band's web site at gutpluckers.com, or find them on Facebook.

ILLUSTRATIONS:

Jessica Helm

Jessica Helm is a graduate of Baylor University with a Bachelor of Fine Arts. She currently lives in Tucson, Arizona where she works as a teacher and freelance artist and designer. She is preparing to begin a master's program in Art Therapy and Family Counseling. As an Art Therapist, she plans to work with children in crisis situations and use her artistic abilities to change lives.

MUSIC:

Merle Travis

Merle Travis (1917-1983) was a country music songwriter, guitarist, and singer. He learned to play banjo and guitar in his native Kentucky, and was noted for his innovative finger picking guitar style. Among his best known songs are Sixteen Tons and Dark as a Dungeon, the tune used here with permission. He was inducted into the Country Music Hall of Fame in 1977. You can learn more about Merle Travis at this unofficial website: ofgc.bizland.com/merle_travis.htm or on Wikipedia.

Learn more about us online:

azhornedtoad.com